A. Aubert, J. Mudge

Directions for Making the Best Composition for the Metals

of Reflecting Telescopes

'

A. Aubert, J. Mudge

Directions for Making the Best Composition for the Metals of Reflecting Telescopes

ISBN/EAN: 9783337294380

Printed in Europe, USA, Canada, Australia, Japan

Cover: Foto ©berggeist007 / pixelio.de

More available books at **www.hansebooks.com**

XVI. *Directions for making the best Composition for the Metals of reflecting Telescopes; together with a Description of the Process for grinding, polishing, and giving the great Speculum the true parabolic Curve. By Mr.* John Mudge; *communicated by* Alexander Aubert, *Esq. F. R. S.*

Read Feb. 27. March 6. and 13. 1777. AS the method of casting, grinding, and polishing the specula of reflecting telescopes, by Mess. MOLYNEUX and HADLEY, which is published in Dr. SMITH'S Optics, is what the workmen have generally followed, and is consequently well known to them; I shall in the following account avoid a repetition of the general directions there given, and only remark upon such parts of that process which I think are essentially defective, and supply them by a method of my own, which, from long and repeated trials, I have found completely to answer the purpose. After, therefore, referring to the above account for the manner of making the gages, patterns, the method of casting, as well as a great many other particulars, I will begin with

The

The beſt compoſition for the ſpecula of reflecting teleſcopes.

The perfection of the metal of which the ſpeculum ſhould be made conſiſts in its hardneſs, whiteneſs, and compactneſs; for upon theſe properties the reflective powers and durability of the ſpeculum depend.   And firſt of the hardneſs and whiteneſs of the metal.   There are various compoſitions recommended in SMITH's Optics, all which have however their ſeveral defects.   Three parts copper and one part and one-fourth of tin will make, he ſays, a very hard white metal; but it is liable to be porous.   This, however, is an imperfection which I ſhall preſently ſhew the method of preventing; but the permanent fault of it, and which I have myſelf experienced, is, that it is not hard enough.   The ſpeculum of a reflecting teleſcope ought to have the utmoſt poſſible hardneſs, compatible with its being operated upon by the tool.

It is to be obſerved, that ever ſo ſmall a quantity of tin added to melted copper deſtroys its perfect malleability, and at the ſame time produces a metal whiter and harder than copper.   As the quantity of tin is increaſed, ſuppoſe to a fifth or fourth part, the metal becomes whiter, ſtill harder, and conſequently more friable.   If the quantity of tin be further increaſed to a third of the whole compoſition, it will then have its utmoſt whiteneſs; but will be rendered at the ſame time ſo exceedingly hard and brittle, that the fineſt waſhed emery upon lead or braſs will not cut it without breaking up its ſurface; and the common blue ſtones uſed in grinding the ſpeculum, will not touch it. Mr. JACKSON (ſome time ſince dead) a mathematical-inſtrument-maker, and a moſt excellent workman, told me, that the tin was increaſed to the above proportion in his metals; but that they were ſo exceedingly hard, that it coſt him an infinite deal of pains, and a journey of two hundred miles, to find out a ſtone of ſufficient hardneſs to cut it, and whoſe texture at the ſame time was fine enough not to injure its ſurface. I have ſeen ſeveral of his finiſhed metals; they were indeed perfectly hard and white; but the kind of ſtone with which he ground them he kept a ſecret.

After many experiments with various proportions of tin and copper, by gradually increaſing the former, I at laſt found that fourteen ounces and an half of grain-tin to two pounds of good Swediſh copper, made a beautiful white and very hard metal; ſo hard indeed, that the ſtones would but barely cut it, and waſhed emery on braſs or tin but juſt grind the ſurface without breaking it up; whereas the proportion of tin being increaſed by the addition of only another half ounce, the former inconvenience

venience immediately took place. This therefore is the *maximum* in point of hardnefs.

Thus much of the two firft confiderations, the hardnefs and whitenefs of the metal; the next, and indeed the moft eiiential, property is its compactnefs, or its being without pores.

This compofition (though complete in the former refpects) was, as well as Dr. SMITH's, fubject every now and then to be porous; fometimes, indeed, I fucceeded in cafting a fingle metal, or perhaps two or three, without this imperfection; at other times, and moft frequently indeed, they were attended with this defect, without my being at all able to form a probable conjecture at the caufe of my fuccefs or difappointment. The pores were fo very fmall that they were not difcoverable when the metal had received a good face and figure upon the hones, nor till the laft and higheft polifh had been given; and then it frequently appeared as if dufted over with millions of microfcopic pores, which were exceedingly prejudicial in two refpects; for firft, they became in time a lodgment for a moifture which tarnifhed the furface; and fecondly, on polifhing the fpeculum, the putty neceffarily rounded off the edges of the pores, fo as to fpoil a great part of the metal, by the lofs of as much light and

fharpnefs

ſharpneſs in the image as there were defective points of reflection in the metal.

Beſides the trouble of a great number of experiments, in order to get rid of this miſchief, and to aſcertain the cauſe to which it was owing, there was this additional inconvenience attending it, *viz.* that the fault was not diſcovered, as was obſerved before, till a great deal of trouble had been ·taken in grinding and even poliſhing the metal, the whole of which was rendered uſeleſs by the mortifying diſcovery of this defect.

I was extricated at laſt from this difficulty, and in ſome meaſure by accident. Having one day made a great number of experiments, and having melted down all the good copper I had or could procure; though puzzled and fatigued, yet not caring to give it up, I recollected that I had ſome metal which was reſerved out of curioſity, and was a part of one the bells of St. Andrew's which had been re-caſt. Expecting, however, very little from this groſs and uncertain compoſition, I was-nevertheleſs determined to ſee what could be made of it by enriching the compoſition with a little freſh tin. Accordingly caſting a metal with it, it turned out perfectly free from pores, and in every reſpect as fine a metal as ever I ſaw.

I could not at firſt conceive to what this ſucceſs was owing; but at laſt I hit upon the real cauſe of that defect,

5   which

which had given me so much embarrassment and trouble during a course of near a hundred experiments, and in consequence thereof fell upon a method which ever after prevented it.

I had hitherto always melted the copper first, and when it was sufficiently fused, I used to add the proportional quantity of tin; and as soon as the two were mixed, and the scoria taken off, the metal was poured into the moulds. I began to consider that putty was calcined tin, and strongly suspected, that the excessive heat which the copper necessarily undergoes before fusion, was sufficient to reduce part of the tin to this state of calcination, which therefore might fly off from the composition in the form of putty, at the time the metal was poured into the flasks.

Upon this idea, after I had furnished myself with some more Swedish copper and grain-tin (both which I had always before used) I melted the copper, and having added the tin as usual to it, cast the whole into an ingot: this was, as I expected, porous. I then melted it again, and as in this mixed state it did not acquire half the heat which was before necessary to melt the copper alone, so it was not sufficient to calcine the tin; the speculum was then perfectly close, and free from this fault; nor did I ever after, in a single instance, meet with the above mentioned imperfection.

All

All that is neceffary, therefore, to be done to procure a metal which fhall be white, as hard as it can be wrought, and perfectly compact, is to melt two pounds of Swedifh copper, and when fo melted, to add fourteen ounces and a half of grain-tin to it; then, having taken off the fcoria, to caft it into an ingot. This metal muft be a fecond time melted to caft the fpeculum; but as it will fufe in this compound ftate with a fmall heat, and therefore will not calcine the tin into putty, it fhould be poured off as foon as it is melted, giving it no more heat than is abfolutely neceffary. It is to be obferved, however, that the fame metal, by frequent melting, lofes fomething of its hardnefs and whitenefs: when this is the cafe, it becomes neceffary to enrich the metal by the addition of a little tin, perhaps in the proportion of half an ounce to a pound. And indeed when the metal is firft made, if inftead of adding the fourteen ounces and a half of tin to the two pounds of melted copper, about one ounce of the tin were to be referved and added to it in the fucceeding melting, before it is caft off into the moulds, the compofition would be the more beautiful, and the grain of it much finer: this I know by experience to be the cafe.

The beft method for giving the melted metal a good furface is this: the moment before it is poured off, throw into the crucible a fpoonful of charcoal-duft; imme-

2   diately

diately after which the metal muft be ftirred with a wooden fpatula, and poured into the moulds.

I wifh I may not be confidered as tedious in the above detail; but as this bufinefs caufed me a great deal of trouble, I was willing to give fome account of the means by which I was freed from this difficulty ever after. Perhaps, indeed, the whole of this procefs may be unneceffary, as many years fince, I communicated this compofition, and I believe at the fame time the method of preventing the pores, to the late Mr. PETER COLLISON, a member of the Royal Society; and likewife two or three years fince, at the defire of my brother, to Mr. MICHELL. Although it be poffible, therefore, that this method is generally known, yet, as I have frequently of late feen fpecula with this defect, and obferved metals of fome of Mr. SHORT's telefcopes which are not quite fo perfect as could be wifhed (though they are all exquifitely figured) I was willing by this publication wholly to remove any future embarraffment of this fort, and to furnifh workmen with an excellent compofition for their metals. And would the Royal Society be pleafed to honour the procefs with a place in their records, I know of no other method fo proper to give this, as well as the following information, a general notoriety.

The metal being caft, there will be no occafion for the complicated apparatus directed by Dr. SMITH, for

grinding

grinding and poliſhing it.   Four tools are all that are
neceſſary, *viz.* the rough grinder to work off the rough
face of the metal; a braſs convex grinder, on which the
metal is to receive its ſpherical figure; a bed of hones
which is to perfect that figure, and to give the metal a
fine ſmooth face; and a concave tool or bruiſer, with
which both the braſs grinder, and the hones are to be
formed.   A poliſher may be conſidered as an additional tool; but as the braſs grinder is uſed for this purpoſe, and its pitchy ſurface is expeditiouſly, and without
difficulty formed by the bruiſer, the apparatus is therefore not enlarged.

Of rough grinding the ſpeculum.

The tool by which the rough ſurface of the metal is
rendered ſmooth and fit for the hones, is beſt made of lead,
ſtiffened with about a fifth or ſixth part of tin. ·This tool
ſhould be at leaſt a third more in diameter than the metal
which is to be ground; and for one of any ſize, not leſs
than an inch thick.   It may be cemented upon a block
of wood, in order to raiſe it higher from the bench.

This leaden tool being caſt, it muſt be fixed in the lathe,
and turned as true as it is poſſible, by the gage, to the
figure of the intended ſpeculum, making a hole or pit in
the

the middle, as a lodgment for the emery, of about an inch diameter for a metal of four inches: when this is done, deep grooves muft be cut acrofs its furface with a graver, in the manner reprefented in fig. 1. Thefe grooves will ferve to lodge the emery, and by their means the tool will cut a great deal fafter. There is no occafion to fear any alteration in the convexity of this tool by working the metal upon it, for the emery will bed itfelf in the lead, and fo far arm the furface of it, that it will preferve its figure and cut the metal very faft. Any kind of low handle, fixed on the back of the metal with foft cement, will be fufficient; but it fhould cover two-thirds of its back to prevent its bending. This way of working will cut the metal fafter, and with more truth, than the method defcribed by Dr. SMITH; for fhould the furface and rough parts be attempted to be ground off by a common grind-ftone by hand, though you did it as near the gage as poffible, yet the metal would be fo much out of truth when applied to the fucceeding tool, that no time would be faved by it. I ufed to employ a common labourer for this purpofe, who foon acquired fuch a dexterity at working upon this tool, that in two hours time he would give a metal of four inches diameter fo good a face and figure as even to fit it for the hones. When all the fand-holes and irregularities on the face of the metal are ground off, and the whole furface is ſmooth and regularly figured, the ſpeculum is then ready for the braſs grinder, and muſt be laid aſide for the preſent.

The manner of forming the braſs-grinding tool.

The following is the method I have always purſued. Procure a round ſtout piece of Hamburgh braſs, at moſt a ſixth part larger than the metal to be poliſhed; and let it be well hammered into a degree of convexity (by the aſſiſtance of the gage) ſuitable to the intended ſpeculum. Having done this, ſcrape and clean the concave ſide ſo thoroughly that it may be well tinned all over; then caſt upon it, after it has been preſſed a proper depth into the ſand, the former compoſition of tin and lead, in ſuch quantity, that it may (for a ſpeculum of four inches diameter) be at leaſt an inch and an half thick, and with a baſe conſiderably broader than the top, in order that it may ſtand firmly upon the bench in the manner hereafter to be deſcribed. This being done, it muſt be fixed and turned in the lathe with great care, and of ſuch a convexity as exactly to ſuit the concave gage, which we ſuppoſe already made. It will be neceſſary to be more careful in forming this than the former tool, and eſpecially that no rings be left from the turning;

I   nor

nor will the fucceeding hone tool require fo much exact-
nefs, as any defects in turning, will, by a method hereafter
mentioned, be eafily remedied; but any inequality or want
of truth in the brafs tool will occafion a great deal of trouble before it can be ground out by the emery. This tool
muft have a hole (fomewhat lefs than that in the metal
to be worked upon it) in the middle, quite through to
the bottom. When this tool is finifhed off in the lathe,
its diameter fhould be one-eighth wider than the metal.

How to form the bed of hones, or the third tool.

Having chofen the kind of hones, and the beft too, of
the fort recommended in SMITH's Optics; they fhould
be cemented in fmall pieces (in a kind of pavement
agreeably to his directions) upon a thick round piece of
marble, or metal made of lead and tin like the former
compofition (which is what I have always ufed) in fuch
a manner, that the lines between the ftones may run
ftraight from one fide to the other; fo that, placing the
teeth of a fine faw in each of thefe divifions, they may
be cleared from one end to the other of the cement
which rifes between the ftones. This bed of hones
fhould be at leaft a fourth part larger than the metal
which is to be ground upon it. The furface of the

R r 2

metal

metal upon which the hone pavement is to be cemented may or may not, as you pleaſe, be turned of a convexity ſuitable to the gage, though I have never taken that trouble. As ſoon as the hones are cemented down, and the joints cleared by the ſaw, this tool muſt be fixed in the lathe, and turned as exactly true to the gage as poſſible; which done, it muſt be laid aſide for the preſent. The next tool to be made is the bruiſer.

The manner of forming the bruiſer, the fourth and laſt tool.

This tool ſhould be likewiſe made of thick ſtout braſs like the former, perfectly ſound, about a quarter of an inch thick, and hammered as near to the gage as poſſible. It ſhould be then ſcraped, cleaned, and tinned on the convex ſide, as the former tool was on the concave, and the ſame thickneſs of lead and tin caſt upon it. The general ſhape of this ſhould differ from the former; for as that increaſed in diameter at the bottom for the ſake of ſtanding firmly, ſo this ſhould be only as broad at bottom as at top, as it is to be uſed occaſionally in both thoſe poſitions. When this tool is fixed in the lathe, and turned off concave to the convex gage with great truth

likewiſe,

likewife, its diameter ought to be the middle fize between the hones and the polifher.

Having with the lathe roughly formed the convex
brafs grinder, the bed of hones, and the concave bruifers,
the convex and concave brafs tools and the metal muft
be wrought alternately and reciprocally upon each other
with fine emery and water, fo as to keep them as nearly
to the fame figure as poffible, in order to which fome
wafhed emery muft be procured. This is beft done by
putting it into a phial, which muft be half filled with
water and well fhaken up, fo that, as it fubfides, the coarfeft
may fall to the bottom firft, and the fineft remain at the
top: and whenever frefh emery is laid on the tools, the
beft method (which we fhould alfo obferve with the
putty in polifhing) will be, to fhake gently the bottle,
and pour out a fmall quantity of the turbid mixture.

Of grinding the fpeculum, the brafs tool, and the bruifer,
together.

All the tools being ready, upon a firm poft in the
middle of a room, you are to begin to grind the brafs convex tool with the bruifer upon it, working the latter crofs-
ways, with ftrokes fometimes acrofs its diameter, at others
a little to the right and left, and always fo fhort that the
bruifers

bruiſers may not paſs above half an inch within the ſurface of the braſs tool either way, ſhifting the bruiſer
round its axis every half dozen ſtrokes or thereabout.
You muſt likewiſe, every now and then, ſhift your own
poſition, by walking round, and working at different ſides
of the braſs tool; at times the ſtrokes ſhould be carried
round and round, but not much over the tool: in ſhort,
they muſt be directed in ſuch a way, and the whole
grinding conducted in ſuch a manner and with ſuch
equability, that every part of both tools may wear
equally. This habit of grinding, as well as the future
one of poliſhing, will be ſoon acquired. When you have
wrought in this manner about a quarter of an hour with
the bruiſer upon the tool, it will be then neceſſary to
change them, and, placing the bruiſer upon its bottom,
to work the convex tool upon that in the ſame manner.

When by working in this equable manner, alternately
with the bruiſer and tool, and occaſionally adding freſh
emery, you have nearly got out all the veſtiges of the
turning tool, and brought them both nearly to a figure,
it will be then time to give the ſame form to the metal.
This muſt be done by now and then grinding it upon
the braſs tool with the ſame kind of emery, taking care
however, by working the two former tools frequently
together, to keep all three exactly in the ſame curve.

The

The beſt kind of handle for the metal is made of lead, a little more than double its thickneſs, and ſomewhat leſs in diameter, of about three pounds weight, with a hole in the middle (for reaſons to be ſhewn hereafter) a little larger than that in the metal: this handle ſhould be cemented on with pitch. The upper edge of this weight muſt be rounded off, that the fingers may not be hurt; and a groove, about the bigneſs of the little finger, be turned round juſt below it, for the more conveniently holding and taking the metal off the tools.

The manner of figuring the metal upon the bones.

When the bruiſer, braſs tool, and metal, are all brought to the ſame figure, and have all a true good ſurface, the next part of the proceſs is to give a correct ſpherical figure and a fine face to the metal, upon the hones. It will be neceſſary to premiſe, however, that the hones ſhould be placed in a veſſel of water, with which they ſhould be quite covered for at leaſt an hour before they are uſed, otherwiſe they will be perpetually altering their figure when the metal comes to be ground upon them. The ſame precaution is alſo neceſſary, if you are called off from the work while you are

grinding

grinding the metal, for if they be ſuffered to grow dry, the ſame inconvenience will ariſe.

In order to give a proper figure to the hones, and exactly ſuitable to that of the braſs tool, bruiſer, and metal, when the hones are fixed down to the block, ſome common flour emery (unwaſhed) with a good deal of water muſt be put upon them, and the bruiſer being placed upon the hones and rubbed thereon with a few ſtrokes and a light hand, the inequalities of the ſtone will be quickly worn off; but as a great deal of mud will be ſuddenly generated, it muſt be waſhed off every quarter of a minute with a great deal of water. By a repetition of this, two or three times, the hones (being of a very ſoft and friable ſubſtance) will be cut down to the figure, without wearing or altering the bruiſer at all. Though this buſineſs may be quickly done, and can be continued but for a few ſtrokes at a time, I need not ſay that it is neceſſary that thoſe ſtrokes be carried in the ſame direction, and with the ſame care, which was obſerved in grinding the former tools together.

As ſoon as the hones have received the general figure of the bruiſer, and all the turning ſtrokes are worn out from them, the emery muſt be carefully waſhed off; in order to which, it will be neceſſary to clear it from the joints with a bruſh under a ſtream of water. The bruiſer

2   and

and metal muſt be likewiſe cleared in the ſame manner, and with equal care, from any lurking particles of emery.

The hones being fixed down to the block, you now begin to work the bruiſer upon them with very cautious, regular, ſhort ſtrokes, forward and backward, to the right and left, turning the axis of the bruiſer in the hand while you move round the hones, by ſhifting your poſition, and walking round the block. Indeed the whole now depends upon a knack in working, which ſhould be conducted nearly in the following manner. Having placed the bruiſer on the centre of the hones, ſlide it in an equable manner forward and backward, with a ſtroke or two directly acroſs the diameter, a little on one ſide, and ſo on the other; then ſhifting your poſition an eighth part round the block, and having turned the bruiſer in your hand about as much, give it a ſtroke or two round and round, but not far over the edges of the hones, and then repeat the croſs ſtrokes as before: thoſe round ſtrokes (which ought not to be above two or three at moſt) are given every time you ſhift your own poſition and that of the metals, previous to the croſs ones, in order to take out any ſtripes either in the hones or bruiſer, which may be ſuppoſed to be occaſioned, by the ſtraight croſs ſtrokes. During the time of working, no mud muſt be ſuffered to collect upon the hones, ſo as to

S ſ deſtroy

deſtroy the perfect contact between the two tools; and therefore they muſt every now and then be waſhed clean by throwing ſome water upon them. When by working in this manner all the emery ſtrokes are ground off from the bruiſer, and it has acquired a good figure and clean ſurface, you may then begin with the metal upon the hones, in the ſame cautious manner, waſhing off the mud as faſt as it collects, though that will be much leſs now than when the bruiſer was ground upon them. Every now and then, however, the bruiſer muſt be rubbed gently and lightly upon the hones, which will as it were, by ſharpening them and preventing too great ſmoothneſs, occaſion them to cut the metal much faſter.

When, after having ſome time cautiouſly wrought in the manner before deſcribed, the hone-pavement has uniformly taken out all the emery ſtrokes, and given a fine face and true figure to the metal, which will be pretty well known by the great equality there is in the feel while you are working, and by which an experienced workman will form a pretty certain judgment; having proceeded thus far, I ſay, you may then try your metal, and judge of its figure by this more certain manner.

Waſh the hone pavement quite clean; then put the metal upon the center of it, and give two or three light

ſtrokes

ftrokes round and round only, not carrying, however, the
edges of the metal much over the hones; this will take out
the order of ftraight ftrokes: then having again wafhed
the hones, and placed the fpeculum upon their center,
with gentle preffure, flide it towards you till its edge be
brought a little over that of the hones, then carry it
quite acrofs the diameter as far the other fide, and having
given the metal a light ftroke or two in this direction,
take it off the tool. The metal being wiped quite dry,
place it upon a table at a little diftance from a window;
ftand yourfelf as near the window, at fome diftance from
the metal, and looking obliquely on its furface, turn it
round its axis, and you will fee at every half turn the
grain given by the laft crofs ftrokes flafh upon your eye
at once over the whole face of the metal. This is as certain a proof of a true fpherical figure as the operofe and
difficult method defcribed in Dr. SMITH's Optics; for as
there is nothing foft or elaftic, either in the metal or in the
hones, this glare is a certain proof of a perfect contact
in every part of the two furfaces; which there could not
be if the fpheres were not both perfect and precifely the
fame.

Indeed there is one accidental circumftance which neceffarily affords its aid in this and every bufinefs of the
like fort; and that is, that a concave and convex furface

ground

ground together, though ever fo irregular at firft, will
(if the working be uniform and proper, confifting, efpe-
ciallv at laft, of crofs ftrokes in every poffible direction
acrofs the diameter) be formed into portions of true and
equal fpheres; had it not been for this lucky neceffity, it
would have been impoffible to have produced that correctnefs which is effential in the fpeculum of a good reflecting telefcope by any mechanic contrivance what-
ever. For when it is confidered, that the errors in reflection are four times as great as in refraction, and that the
leaft defect in figure is magnified by the powers of the
inftrument, any thing fhort of perfection in the figure
of the fpeculum would be evidently perceived by a want
of diftinctnefs in the performance.

L muft not, however, quit this article without obferving, that I all along fuppofe, both in forming the tools
and at laft figuring the metal (and indeed the fame muft
be obferved in the future procefs of polifhing) that no
kind of preffure is ufed that may endanger the bending
or irregularly grinding them; they fhould therefore be
held with a light hand, and loofely between the fingers,
and the motion given fhould be in a horizontal direction,
with no more preffure than their own dead weight.

Having now finifhed the metal on the hones, and rendered it both in point of figure and furface fit for the

laft

laft and moft effential procefs, *viz.* that of polifhing, I will defcribe it in the beft manner I can; though many little circumftances which will be unavoidably omitted (and which at the fame time are frequently effential to the fuccefs of a mechanic procefs) can only be fupplied by actual experience.

The polifhing the fpeculum is the moft difficult and effential part of the whole procefs; for every experienced workman knows, to his vexation, that the moft trifling error here will be fufficient to fpoil the figure of his metal, and render all his preceding caution ufelefs. I have, however, difcovered a method which I fhall explain, not only of giving the metal a parabolic figure, but alfo of recovering it when it happens to be injured; both to be effected in the act of polifhing, and the former as certainly as the fpherical figure is given upon the hones. Indeed, if we confider rightly, polifhing will be perceived to be but a kind of grinding with a finer order of ftrokes, and with a powder infinitely finer than was before ufed in what is commonly called the grinding. But before I defcribe this method, which was the refult of many years experience, I will take the liberty of making fome few ftrictures on that of Meff. HADLEY and MOLYNEUX, which is followed by the generality of workmen.

Firſt, then, the tool itſelf uſed by them for poliſhing the metal, is formed with infinite difficulty. The
firſt deſcribed poliſher is directed to be made by covering
the tool with ſarcenet, which is to be ſaturated with a
ſolution of pitch in ſpirit of wine, by ſucceſſive applications of it with a bruſh, till it is covered, and by the evaporation of the ſpirit of wine filled with this extract of
pitch; the ſurface is then to be worked down and finiſhed
with the bruiſer. This is all very eaſy in imagination;
but whoever has uſed this method (which I have myſelf
unſucceſsfully ſeveral times) muſt have found it attended
with infinite labour, and at laſt the buſineſs done in a
very unſatisfactory manner; for the pitch by this proceſs
will be deprived of an eſſential part of its compoſition.
The ſpirit of wine diſſolves none but the reſinous parts
of its ſubſtance, which is hard and untractable; and if
you uſe ſoap or ſpirit of wine to ſoften or diſſolve it, it
will equally affect the whole ſurface, the lower as well as
higher parts of it. And ſuppoſe that with infinite labour
with the bruiſer, it is at laſt reduced to a fine uniform
ſurface, it is nevertheleſs too hard ever to give a good
poliſh with that luſtre which is always ſeen in Mr.
SHORT's, and indeed all other good metals. Nor will it
give a good ſpherical figure; for a perfect ſphere is
formed, as I obſerved before, by that intimate accommodation

dation arifing from the wear and yielding of both tool and metal; whereas in this method, there is fuch a ftubbornnefs in the polifher, that the figure of the metal, good or bad, muft depend upon the truth of the former, which is very feldom perfect.

If the polifher be made in the fecond manner propofed, by ftraining the pitch through an outer covering, which is afterwards to be ftripped off, the fuperficies of pitch and farcenet is fo very thin, that the putty, working into them, forms a furface hard and untractable, fo that it is impoffible to give the fpeculum a fine polifh. Accordingly all thofe metals which are wrought that way have an order of fcratches inftead of polifh, difcovering itfelf by a greyifh vifible furface. Befides, fuppofing this tool perfectly finifhed, and anfwering its purpofe ever fo well, it is impoffible it can produce in the fpeculum any other than a fpherical figure; and indeed nothing elfe is expected from this method, as very evidently appears by the experiment recommended to afcertain the truth of the figure. You are directed to place a fmall luminous object in the center of the fphere of which the metal is a fegment; and then having adjufted an eye-glafs at the diftance of its own focal length from the object, and fo fituated that the image of the object formed by the fpeculum may be vifible to the eye, you are to judge of the

3   perfect

perfect figure of the metal by the ſharpneſs and diſtinctneſs with which the image appears. From hence it is
very evident, that as the object and image are both diſtant
from the metal by exactly its radius, nothing but a true
ſpherical figure of the ſpeculum can produce a ſharp
diſtinct image; and that the image could not be diſtinct
if the figure of the ſpeculum were parabolic. Conſequently, if the ſame ſpeculum uſed in a teleſcope were
to receive parallel rays, there would neceſſarily be a conſiderable aberration produced, and a conſequent imperfection in the image. Accordingly, there never was a
good teleſcope made in this manner; for if the number
of degrees, or the portion of the ſphere of which the
great metal is a part, were as conſiderable as it ought to
be, or as great as Mr. SHORT allowed in his metal, the
inſtrument would bear but a very low charge, unleſs a
great part of the circumference of the metal were cut off
by an aperture, and the ill effects of the aberration by
that means in ſome meaſure prevented.

If ever a finiſhed metal turned out without this defect,
and has been found perfectly ſharp and diſtinct, it muſt
have been owing to an accidental parabolic tendency, no
ways the natural reſult of the proceſs, and therefore quite
unexpected, and moſt probably unknown, to the workman.

Without

Without enlarging, therefore, on the difficulty of the above procefs, and the impoffibility of giving the fpeculum the correctnefs and the kind of figure effentially neceffary to a good telefcope, I will defcribe (by way of introduction to the fucceeding directions) the fteps by which I was led to a certain and eafy method of giving a proper and correct parabolic figure to the metal, even though it came off imperfect from the hones, and an exquifite polifh at the fame time.

Having made many efforts in the former method, which by no means pleafed me for the reafons above-mentioned; and having obferved, from fome of Mr. short's telefcopes which fell into my hands, that the high luftre of the polifh could never have been produced in the manner above defcribed, but by fome fofter and more tender fubftance; and at the fame time recollecting, that Sir isaac newton had given an account in his Optics of his having finifhed fome metals, and confiderably mended the object glafs of a refractor, by working both upon a tool whofe furface had been covered with common pitch about the thicknefs of a groat; reflecting, I fay, upon thefe matters (coarfe and uncertain as this method appeared at firft fight) I was determined to try whether I could not get rid of my embarraffment, by a mode of operation fomewhat fimilar. Accordingly, fhortening

Dr. SMITH's procefs, I made a fet of tools in the manner
before defcribed, except that I was obliged to make fome
fubfequent alteration in the polifher which I fhall pre-
fently defcribe.   Having given a good fpherical figure
to the brafs tool and the bruifer, and likewife to the
metal upon the hones, and made the brafs convex tool fo
hot as juft not to hurt the finger, I tied a lump of common pitch (which fhould be neither too hard nor too
foft) in a rag, and holding it in a pair of tongs over a ftill
fire where there was no rifing duft, till it was ready to
ftrain through the linen, I caufed it to drop upon the
feveral parts of the convex tool, till I fuppofed it would
cover the whole furface about double the thicknefs of a
fhilling; then fpreading the pitch as equally as I could, I
fuffered the polifher (by which name I fhall for the
future call this tool) to grow quite cold.   I then warmed
the bruifer fo hot as almoft to burn my fingers, and having fixed it to the bench with its face upwards, I fud-
denly placed the polifher upon it, and quickly flid it off;
by this means rendering the furface of the pitch more
equal.   The pitch is then to be wiped off from the bruifer
with a little tow; and by touching the furface with a
tallow candle, and wiping it a fecond time, it will be then
perfectly clean and fit for a fecond procefs of the fame
fort, which muft again be performed as quickly as poffible;

and

and this is ordinarily fufficient to give a general figure to the furface of the pitch. The bruiler and polifher are then fuffered to grow perfectly cold, when the pitch, confidering what has been taken off, will be about the thicknefs of a fhilling.

It is however here neceffary to obferve, that the pitch fhould be neither very hard and refinous, nor too foft; if the former, it will be fo untractable as not to work kindly; and if too foft, it will in working alter its figure fafter than the metal, and too readily fit itfelf to the ir-regularity of its figure, if it have any. When both tools were perfectly cold, I gave the polifher a gentle warmth, and then fixed the bruifer to the block with its face up-wards; and (having with a large camel's-hair brufh fpread over the face of the polifher a little water and foap, to prevent fticking) with fhort, ftraight, and round ftrokes I worked it upon the bruifer, every now and then adding a little more water and foap, till the pitch upon the polifher had a fine furface, and the true form of the bruifer; and this I continued to do till they both grew perfectly cold together: in this manner the polifher was perfectly formed in about a quarter of an hour. But here a difficulty arofe: when I begun to polifh the metal, I found that the edge of the hole in the metal collected the pitch towards the middle of the polifher; and

though

though in this method of working I could give an ex-
quiſite poliſh, as the putty lodged itſelf in the pitch
exceedingly well, yet the figure of the metal was injured
in the middle, nor did indeed the work go on with that
equability which is the inſeparable attendant on a good
figure. In order to obviate this difficulty, I caſt ſome
metals with a continued face, the holes not going quite
through, within perhaps the thickneſs of a ſix-pence.
I finiſhed two or three metals of this ſort, and the work
promiſed and went on very well; but when I came to open
the holes, which I did with the utmoſt caution, I found
the metals ſhort of perfection; which I attributed to an
alteration of the figure from the removal of even that ſmall
portion of metal after the ſpeculum had been finiſhed.
This I do ſuppoſe was in ſome meaſure the reaſon why I
ſpoiled a very diſtinct and perfect two-foot metal, which
bore a charge of two hundred times, only by opening
the ſharp part of the edge of the hole, becauſe I thought
it bounded the field: ſo eſſentially neceſſary is an exqui-
ſite correctneſs of figure in the ſpeculum of a perfect
reflector.

This experiment not ſucceeding, inſtead of caſting the
metal without a hole, I made one quite through the
middle of the poliſher, a little leſs than that in the ſpeculum. This perfectly anſwered the purpoſe; no more
incon-

inconvenience arofe from the gathering of the pitch (for it had now no greater tendency to collect at the center than the fides) and I finifhed feveral metals fucceffively, excellent both in point of figure and polifh ; one of thofe of two inches diameter and 7,5 focal length, bore a charge of fixty times and upwards, which when mounted in a telefcope I gave to my brother. This telefcope underwent Mr. SHORT's examination, who was pleafed to remark only, that he thought he had made one more diftinct.

I muft obferve, that in this method of working the polifhing goes on in an agreeable, uniform, and fmooth manner; and that the fmall degree of yielding in the pitch (which is actually not more than the wearing of the metal) produces that mutual accommodation of furfaces fo neceffary to a true figure. In the beginning of the polifh, and indeed for fome time during the progrefs of it (always remembering now and then to move the metal round its axis) I worked round and round, not far from and always equally diftant from the center, except that every time, previous to the fhifting the metal on its axis, I ufed a crofs ftroke or two; and when the polifh was nearly compleated, I moftly ufed crofs ftrokes, giving a round ftroke or two likewife every time I turned the metal on its axis. I obferved in this method of working, that the metal always polifhed fafteft in the middle;

infomuch,

infomuch, that half or two-thirds of it would be completely poliſhed when the circumference of it was fcarcely
touched by the tool. Obferving this in fome of the firſt
metals, and not confidering that this way of poliſhing
was in fact a fpecies of grinding, and as perfect as that
upon the hones, I went on reluctantly with the work,
almoſt defpairing of being able to produce a good
figure.   However, I always found myfelf agreeably
deceived; for when the poliſh was extended to the
edge, or within the tenth of an inch of it, I almoſt
conſtantly found the figure good, and the performance
of the metal very diſtinct. But this fame circumftance of
apparent defect in the metals, was in fact that to which
their perfection was owing; for they all, contrary to my
expectation, turned out parabolic.   However I did not
for a great while know any certain way of giving that
degree of parabolic tendency which was juſt neceſſary,
and which will be defcribed hereafter.   It was a long
time before I got rid of my prejudice againſt this apparent imperfection in the procefs, or could reconcile myfelf to the irregular manner in which the poliſh proceeded; for I looked upon it as a certain fource of error,
and notwithſtanding I faw it eventually fucceed, yet
whenever I chanced to find that a metal, when firſt applied to the poliſher, took the poliſh equally all over,

and

and confequently the whole bufinefs did not take up above ten minutes; under thofe circumftances, I fay, I always ufed to pleafe myfelf with the expectation of a correct figure, at leaft as much fo as the metal had received from the hones, where the furface was but juft and equally taken off by the putty; but in this I conftantly found myfelf deceived, and the metal turned out good for nothing. In fhort, at this time, though I fpeculatively knew that a parabolic figure was neceffary to a perfect image, I yet confidered it as of little practical confequence.

From the foregoing experiments, and a number of fucceeding trials, I at length difcovered a certain way of giving a correct parabolic figure, and an exquifite polifh at the fame time. This, which I have ftrong reafons to believe was Mr. SHORT's method, I will now defcribe in as few words as I can.

How to polifh the fpeculum.

It is firft neceffary to obferve, that, in order to avoid the detrimental intrufions of any particles of emery, it would not be right to polifh in the fame room where the metal and tools were ground, nor in the fame cloaths which were worn in the former procefs; at leaft it would

L   be

be neceſſary to keep the bench quite wet, to prevent any duſt from riſing.

Having then made the poliſher by coating the braſs convex tool equally with pitch, which we ſuppoſe ſmoothed and finiſhed with the braſs tool in the manner before deſcribed, and which is a very eaſy procefs, the whole operation is begun and finiſhed in the following manner.

The leaden weight or handle upon the back of the metal ſhould be divided into eight parts, by ſo many deep ſtrokes of a graver upon the upper ſurface of the lead, marking each ſtroke with the numbers 1, 2, 3, 4, and ſo on, that the turns of the metal in the hand may be known to be uniform and regular.

To prevent any miſchief from coarſe particles of putty, I always waſh it immediately before uſing. In order to this, put about half an ounce of putty into an ounce phial, and fill it two-thirds with water; then having ſhaken the whole, let the putty ſubſide, and ſtop the bottle with a cork.

In a tea-cup with a little water, there ſhould be a full-ſized camel's-hair bruſh, and a piece of dry clean ſoap in a galley-pot: a ſoft piece of ſpunge will alſo be neceſſary. Theſe, as well as the metal bruiſer and poliſher, ſhould be conſtantly covered from duſt.

The

The polisher being fixed down, and the camel's-hair brush, being first wetted and rubbed a little over the soap, let every part of the tool be brushed over therewith; then work the bruiser with short, straight, and round strokes, lightly upon the tool, and continue to do so, now and then turning it, till the polisher have a good face, and be fit for the metal. Then having shaken up the putty in the phial, and touched the polisher in five or six places with the cork wetted with that and the water, place the bruiser upon the tool, and give a few strokes upon the putty to rub down any gritty particles; after which, having removed it, work the metal lightly upon the polisher round and round, carrying the edges of the speculum, however, not quite half an inch over the edge of the tool, and now and then with a crofs stroke.

The first putty, and indeed all the succeeding applications of it, should be wrought with a considerable while; for if time be not given for the putty to bed itself in the pitch, and any quantity of it lie loose upon the polisher, it will accumulate into knobs, which will injure the figure of the metal: and therefore as often as ever such knobs arise, they must be carefully scraped off with the point of a penknife, and the loose stuff taken away with the brush. After the putty is well wrought into the pitch, some more may be added in the same manner, but never much at a time, and always remembering to work upon it firſt with the bruiſer, for fear any gritty particles may find their way upon the poliſher. If the bruiſer be apt to ſtick, and do not ſlide ſmoothly upon the pitch, the ſurface of either tool may be occaſionally bruſhed over with the ſoap and water, but it muſt be remembered that the wet bruſh muſt be but lightly rubbed upon the ſoap.

In the beginning of this proceſs little effect is produced, and the metal does not ſeem to poliſh faſt, in ſome meaſure owing to its taking the poliſh in the middle, and perhaps becauſe neither that nor the bruiſer move evenly upon the poliſher: but a little perſeverance will bring the whole into a good temper of working; and, when the pitch is well defended by the coating of the putty, the proceſs will advance apace, and the former acquiring poſſibly ſome little warmth, the metal moves more agreeably over it, with an uniform and regular friction. All this while the metal muſt have no more preſſure than that which it derives from its own weight and that of the handle; and the poliſher muſt never be ſuffered to grow dry, but, as often as it has any tendency to do ſo, the edges of it muſt be moiſtened with the hair-pencil; and now and then, even when freſh putty is not laid on, the ſurface of the poliſher ſhould be touched with the bruſh to keep it moiſt.

4   When

When the polifh of the metal nearly reaches the edge (for it always, as I faid before, begins in the middle) you muft alter your method of working; for now the round ftrokes muft be gradually altered for the fhort and ftraight ones. Suppofing then you are juft beginning to alter them; after having put on frefh putty, and gently rubbed it with two or three ftrokes of the bruifer, you place the metal on the tool, and after a ftroke or two round and round, give it a few forward and backward, and from fide to fide, but with the edges very little over the tool; then having turned the metal one-eighth round in your hand, and having moved yourfelf as much round the block (which muft be remembered throughout the whole procefs) you go on again with a ftroke or two round, to lead you only to the crofs ftrokes, which are now to be principally ufed, and with more boldnefs. After this has been done fome time, the metal will begin to move ftiffly as the friction now increafes, and the fpeculum polifhes very beautifully and faft; and the whole furface of the polifhing tool will be equally covered over with a fine metallic bronze. The tool even now muft not be fuffered to become dry; a fingle round ftroke in each of your ftations and turnings of the metal will be fufficient, and the reft muft all be crofs ones, for we are completing a circular figure. You muft now be very

diligent,

diligent, for the poliſher drying, and the friction increaſing very faſt, the buſineſs of the ſpherical figure is nearly at an end. As the metal wears much, its ſurface muſt be now and then cleaned, with a piece of ſhammy leather, from the black ſtuff which collects upon it; and the poliſher likewiſe from the ſame matter, with a ſoft piece of wet ſponge. You will now be able to judge of the perfect ſpherical figure of the metal and tool, when there is a perfect correſpondence between the ſurfaces, by the fine equable feel there is in working, which is totally free from all jerks and inequalities. Having proceeded thus far, you may put the laſt finiſhing to this figure of the metal by bold croſs ſtrokes, only three or four in the directions of each of the eight diameters, turning the metal at the ſame time: this muſt be done quickly, for it ought, in this part of the proceſs particularly, to be remembered, that, if you permit the tool to grow quite dry, you will never be able, with all your force, to ſeparate that and the metal, without deſtroying the poliſher by heat.

The metal has now a beautiful poliſh and a true ſpherical figure, but will by no means make a ſharp diſtinct image in the teleſcope: for the ſpeculum (if it be tried in the manner hereafter recommended) will not be found to make parallel rays converge without great aberration;

aberration; indeed the deviation will be fo great, as to be very fenfibly perceived by a great indiftinctnefs in the image.

How to give the parabolic figure to the metal.

In order then to give the fpeculum the laft and finifh-ing figure, which is done by a few ftrokes, it muft be particularly remarked, that by working the metal round and round, the fphere of the polifher by this means growing lefs, it wears fafteft in the middle: and as a fegment of a fphere may become parabolic, by open-ing the extremes gradually from within outwards, fo it may be equally well done by increafing the curva-ture in the middle, in a certain ratio, from without in-wards.

Suppofing then the metal to be now truly fpherical, ftop the hole in the polifher, by forcing a cork into it underneath, about an inch, fo that it do not reach quite to the furface; and having wafhed off any mud that may be on the furface of the tool with a wet foft piece of fponge, whilft the furface of it is a little moift, place the center of the metal upon the middle of the polifher; then having, with the wet brufh, lodged as much water round the edge of the metal as the projecting edge will hold,

hold, fill the hole of the metal and its handle with water, to prevent the evaporation of the moifture, and the confequent adhefion between the fpeculum and polifher, and let the whole reft in this ftate two or three hours: this will produce an intimate contact between the two, and by parting, with any degree of warmth they may have acquired by the vicinity of the operator, they will grow perfectly cold together.

By this time you may pufh out the cork from the polifher, to difcharge the water, and give the metal the parabolic figure in the following manner.

Move the metal gently and flowly at firft, a very little round the centre of the polifher (indeed after this reft it will move ftiffly) then increafing by degrees the diameter of thefe ftrokes, and turning the metal frequently round its axis, give it a larger circular motion, and this without any preffure but its own weight, and holding it loofely between the fingers: this manner of working may fafely be continued about two minutes, moving yourfelf as ufual round the block, and carrying the round ftrokes in their increafed and largeft ftate, not more than will move the edge of the metal half an inch or five-eighths over the tool. The fpeculum muft not all this while be taken off from the polifher; and confequently no frefh putty can be added.

added. It will not be safe to continue this motion longer than the time above-mentioned; for if the parabolic tendency be carried the least too far, it will be impossible to recover a true figure of that kind but by going through the whole process for the spherical one in the manner before described, by the cross strokes upon the polisher, which takes a great deal of time. However, when there is occasion, it may be done; and I have myself several times recovered the circular figure, when I had inadvertently gone too far with the parabolic; and ultimately finished the metal on the polisher without the use of the hones.

To try the true figure of the metal.

It will now be proper to try the figure of the speculum, and that is always best done by placing it in the telescope it is intended for. In order to this, I use the instrument as a kind of microscope, placing the object, however, at such a distance that the rays may be nearly parallel. At about twenty yards a watch-paper, or some such object, on which there are some very fine hair strokes of a graver, is fixed up. The lead must be then taken off from the back of the speculum; which is best done by placing the edge of a knife at the junction of the lead and metal, when, by striking the back of it with a

slight

flight blow, the pitch immediately ſeparates, and the handle drops off; the remaining pitch may be ſcraped off with a knife, taking care that none of the duſt ſtick to the poliſhed face of the metaL

Having placed the ſpeculum in the cell of the tube, and directed the inſtrument to the object, make an annular kind of diaphragm with card-paper, ſo as to cover a circular portion of the middle part of the metal between the hole and the circumference, equal in breadth to about an eighth part of the diameter of the ſpeculum: this paper ring ſhould be fixed in the mouth of the teleſcope, and remain ſo during the whole experiment, for the part of the metal covered by it is ſuppoſed to be perfect, and therefore unemployed.

There muſt likewiſe be two other circular pieces of card-paper cut out, of ſuch ſizes, that one may cover the center of the metal by completely filling the hole in the laſt deſcribed annular piece; and the other, ſuch a round piece as ſhall exactly fit into the tube, and ſo broad as that the inner edge may juſt touch the outward circumference of the middle annular piece. It would be convenient to have theſe two laſt pieces ſo fixed to an axis that they may be put in their places, or removed from thence ſo eaſily as not to diſplace or ſhake the

3   inſtrument.

inftrument. All thefe pieces therefore together will completely fhut up the mouth of the telefcope.

Let the round piece which covers the center of the metal, or that which has no hole in it, be removed; and, by a nice adjuftment of the fcrew, let the image (which is now formed by the center of the mirror) be made as fharp and diftinct as poffible. This being done, every thing elfe remaining at reft, replace the central piece, and remove the outfide annular one, by which means the circumference only of the fpeculum will be expofed, and the image now formed will be from the rays reflected from the outfide of the metal. If there be no occafion to move the fcrew and little metal, and the two images formed by thefe two portions of the metal be perfectly fharp and equally diftinct, the fpeculum is perfect, and of the true parabolic curve; or at leaft the errors of the great and little fpeculum, if there be any, are corrected by each other.

If, on the contrary, under the laft circumftance, the image from the outfide of the metal fhould not be diftinct, and it fhould become neceffary, in order to make it fo, that the little fpeculum be brought nearer, it is plain that the metal is not yet brought to the parabolic figure; but if, on the other hand, in order to procure diftinctnefs, you be obliged to move the little fpeculum farther off, then the figure of the great fpeculum has been carried beyond the parabolic, and hath aſſumed an hyperbolic form. When the latter is the caſe, the circular figure of the metal muſt be recovered (after having fixed on the handle with ſoft pitch) by bold croſs ſtrokes upon the poliſher, finiſhing it again in the manner above deſcribed. If the ſpeculum be not yet brought to the parabolic form, it muſt cautiouſly have a few more round ſtrokes upon the poliſher; indeed a very few of them in the manner before deſcribed make in effect a greater difference in the ſpeculum than would be at firſt imagined. If a metal of a true ſpherical figure were to be tried in the above mentioned manner in the teleſcope (which I have frequently done) the difference of the foci of the two ſegments of the metal would be ſo conſiderable, as to require two or three turns of the ſcrew to adjuſt them; ſo very great is the aberration of a ſpherical figure of the ſpeculum, and ſo improper to procure that ſharpneſs and preciſion ſo neceſſary to a good reflecting teleſcope.

This is by no means the caſe with the object glaſſes of refractors; for beſides that they are in fact never ſo diſtinct as well-finiſhed reflectors, the apertures of them are ſo exceedingly ſmall, compared to the latter, and the number of degrees employed ſo very ſmall, that the inconvenience of a ſpherical figure is not ſo much perceived. Accordingly we obſerve in the generality of

reflectors

reflectors (whofe fpecula, unlefs by accident, are always fpherical) that the only true rays which form the diftinct image arife from the middle of the metal: and unlefs the defect be remedied by a confiderable aperture, which deftroys much light, the falfe reflection from the infide of the metal produces a greyifh kind of hazinefs, which is never feen in Mr. SHORT's or indeed in any good telefcopes.

Suppofing that the two foci of the different parts of the metal perfectly coincide, and that, by the union of them when the apertures are removed, the telefcope fhews the objects very fharp and diftinct, you are not however even then to conclude that the inftrument is not capable of farther improvement; for you will perceive a fenfible difference in the fharpnefs of the image, under different pofitions of the great fpeculum with refpect to the little one, by turning round the great metal in its cell, and oppofing different parts of it to different parts of the little metal, correcting by this means the error of one by the other. This attempt fhould be perfevered in for fome time, turning round the great fpeculum about one-fixteenth at a time, and carefully obferving the moft diftinct fituation each time the eye-piece is fcrewed on: when, by trying and turning the great metal all round, the diftincteft pofition is difcovered, the upper part of the

metal

metal ſhould be marked with a black ſtroke, in order that it may always be lodged in the cell in the ſame poſition. This is the method Mr. SHORT always uſed; and the caution is of ſo much conſequence, that he thought it neceſſary to mention it very particularly in his printed directions for the uſe of the inſtrument.

And farther, Mr. SHORT frequently corrected the errors of the great by the little metal in another way. If the great ſpeculum did not anſwer quite well in the teleſcope, he cured that defect ſometimes by trying the effect of ſeveral metals ſucceſſively, by this means correcting the errors of one by the other; for in ſeveral of his teleſcopes which have paſſed through my hands, when the ſizes and powers have been the ſame, I have found that the great metals, though very diſtinct in their proper teleſcopes, yet have, when taken out and changed from one to the other, ſpoiled both teleſcopes, rendering them exceedingly indiſtinct, which could ariſe from no other circumſtance. For this reaſon I ſuppoſe it was, that he kept, ready finiſhed, a great many large metals of the ſame focal length, ſo that, when he wanted to mount a teleſcope, he might from a great choice, be able to combine thoſe metals which ſuited each other beſt. I am ſtrongly inclined to believe this was the caſe, not only from the above obſervation, but becauſe

he

he shewed me himself a box of finished metals, in which
I am sure there were a dozen and a half of the same focal
length.

To return: a little use in working will make the whole
of the procefs of grinding and polishing very eafy and
certain; for though I have endeavoured to be as particular as I can (I am almoft afraid too much fo) it is yet
scarcely poffible to fupply a want of dexterity, arifing
from habit only, by the moft laboured and minute de-
fcription. And though the above account may appear
irkfome to the reader, as it lies cold before the eye, I am
very fure, whoever attempts to make the inftrument,
will not complain of it as tedioufly particular.

I will, however, farther remark, that when the metal
begins to move ftiffly upon the polifher, and particularly
when the figure is almoft brought to the parabolic form,
it will be neceffary to fix the elbows againft the fides, in
order to give momentum and equability to the motion
of the hand by that of the whole body.

The fame polifher will ferve for feveral metals, if it
be fomewhat warmed when you begin to ufe it.

There is another circumftance, and a material one.
too, which muft not be omitted; it is this. For the very
fame reafon that the pitch fhould not be too hard or foft,
the work will not proceed well in the heat of fummer,

or the cold of winter: in the latter, it may be poſſible to remedy the defect by having the room warmed with a ſtove; and in the ſummer, the other inconvenience may perhaps be avoided by uſing a harder kind of pitch; but I much doubt in either caſe whether the work will go on ſo kindly: I have myſelf always wrought in ſpring and autumn.

The proceſs of poliſhing, and indeed grinding upon the hones, will not go on ſo well if it be not continued uninterruptedly from beginning to end; for if the work of either kind be left but for a quarter of an hour, and you then return to it again, it will be ſome time before the tool and metal can get into a kindly way of working; and till they do, you are hurting what was done before.

I have all along ſuppoſed that the metal we have been working was about four inches diameter: if it be either larger or ſmaller, the ſizes of the hones, bruiſer, and poliſher, muſt be proportionably different. I never find any ill conſequence ariſing from the different expanſion from heat and cold in any of the tools, though they be made of different metals and ſubſtances, unleſs the inconvenience, occaſioned by the interruption before hinted at, be thought to reſult from thence; for the alteration produced in the ſurface of the ſpeculum, both by grinding and poliſhing, is ſo much quicker than any that can be ſuppoſed to ariſe

from

from the former caufe, that it is never attended with any practical confequence.

Magnifying very minute objects, and particularly reading at a diftance, have been generally confidered as the fureft tefts of the goodnefs of a telefcope; and indeed when the page is placed at a great diftance, fo that the letters fubtend but a very fmall angle at the eye, if then they appear with great precifion and fharpnefs, it is moft probable that the inftrument is a good one. But we are, neverthelefs, fometimes apt to be deceived by this method; nor is it always poffible to determine upon the different merits of two inftruments of equal power, by this mode of examination; for when the letters are removed to the utmoft extent of the powers of the two inftruments, the eye is apt to be prejudiced by the imagination. If two or three words can be here and there made out, all the reft are gueffed at by the fenfe; infomuch that an obferver, zealous for the honour of his inftrument, is very apt to deceive himfelf in fpite of his intentions. The furer teft is by figures, where you can procure no aid from this fort of deception. In order to examine my reflecting telefcopes, I made upon a piece of copper and on a black ground, fix lines confifting of about twelve pieces of gold figures, and each line of figures differing in magnitude, from the fmalleft that could be diftinctly made to thofe

off

of about two-tenths of an inch long; moreover, the figures in the ſeveral lines were differently diſpoſed, and the ſum of each line alſo differed. It is evident that by this method all gueſs is precluded; and that of two inſtruments, of the ſame powers, that which can make out the leaſt order of figures, which will be known by the ſum, is the beſt teleſcope. Such a plate I cauſed to be fixed up for experiments againſt the top of a ſteeple, about three hundred yards North. of my houſe; and it will ſerve to give ſome idea of the diſtinctneſs with which very ſmall figures could be made out at that diſtance, by ſaying, that in a clear ſtate of the air, and with the Sun behind me, with a teleſcope of eighteen inches focal length, which Count BRUHL did me the honour to accept and now has in his poſſeſſion, I have ſeen the legs of a ſmall fly, and the ſhadows of them, with great preciſion and exactneſs.

I cannot conclude without indulging myſelf in an obſervation on the amazing ſagacity of Sir ISAAC NEWTON in every ſubject upon which he thought fit to employ his attention. It was he who firſt propoſed, and indeed practiſed, the poliſhing with pitch; a ſubſtance which at firſt ſight perhaps every one but himſelf would have thought very improper, from its ſoftneſs, to produce that correctneſs of figure ſo neceſſary upon theſe occaſions; and yet I do believe, that it is the only ſubſtance in nature

that

that is perfectly well calculated for the purpose; for at the fame time that it is foft enough to fuffer the putty to lodge very freely on its furface, and for that reafon to give a moft tender and delicate polifh; it is likewife totally inelaftic, and therefore never, from that principle, fuffers any alteration in the figure you give it. If the firft makers of the inftrument, therefore, had given proper credit to, or had fimply followed the hint Sir ISAAC gave, it would have faved them infinite trouble, and they would have produced much better inftruments; but the pretended refinement, of drawing a tincture from pitch with fpirits of wine, affords you only the re-finous, hard, and untractable part of the pitch, divefted of all that part of its original fubftance which is neceffary to give it that accommodating pliability in which its excellence confifts.

It is needlefs to fwell this account with a detail of the procefs for polifhing the little fpeculum, as it muft be conducted in the fame manner which has been already defcribed in that of the large one; only obferving, that as the little metal has an uninterrupted face, without a hole, fo there is no occafion for one in the polifher; and likewife that, as a fpherical figure is all that need here be practically attempted, fo the difficulty in finifhing is infi-nitely fhort of that of the other.

As

As it is always neceſſary to folder to the back of the little ſpeculum a piece of braſs, as a fixture for the ſcrew to adjuſt its axis, I ſhall juſt hint a ſafe and neat method of doing it, which may be very uſeful to the optical or mathematical inſtrument-maker upon other occaſions. Having cleaned the parts to be foldered very well, cut out a piece of tin-foil the exact ſize of them; then dip a feather into a pretty ſtrong ſolution of *ſal ammoniac* in water, and rub it over the ſurfaces to be foldered; after which place the tin-foil between them as faſt as you can (for the air will quickly corrode their ſurfaces ſo as to prevent the folder taking) and give the whole a gradual and ſufficient heat to melt the tin. If the joints to be foldered have been made very flat, they will not be thicker than a hair: though the ſurfaces be ever ſo extenſive, the foldering may be conducted in the ſame manner, only that care muſt be taken, by general preſſure, to keep them cloſe together. In this manner, for inſtance, a ſilver graduated plate may be foldered on to the braſs limb of a quadrant, ſo as not to be diſcernable by any thing but the different colour of the metals. This method was communicated to me by the late Mr. JACKSON, who during his life kept it a ſecret, as he uſed it in the conſtruction of his quadrants, and is, I believe, not as yet known to any workman.

In

In the annexed plate are figured the shape of the leaden tool for rough-grinding; the hones; and the apparatus to be applied to the mouth of the telescope, to ascertain the true figure of the speculum.

POSTSCRIPT.

It was some time after I had written the above account that I saw Mr. SHORT's method of polishing object glasses for refracting telescopes, which is published in the Transactions. By that paper I find that what I before strongly suspected is really the case, viz. that he knew how well pitch was calculated for purposes of this kind. Only it may be remarked, that as glass is much harder, polishes much slower, and consequently does not wear away and alter its figure so soon as the metal of which the speculum is made; and as at the same time (on account of the very small apertures allowed to telescopes of this sort) nothing more than a spherical figure is proposed; he is therefore obliged to use pitch in a hard, friable, and stubborn state: whereas, considering the delicate substance of the metal speculum, and the figure intended to be given to it, the soft pitch of the common sort, by suffering the putty to bed itself in its substance,

produces

produces the moſt beautiful poliſh; and by its pliability is better calculated for that mutual accommodation between poliſher and metal, ſo neceſſary to the figure propoſed.

EXPLANATION OF THE FIGURES.

Fig. 1. The grinder for working off the rough face of the metal; the black ſtrokes repreſent deep grooves made with a graver.

Fig. 2. The bed of hones, which is to complete the ſpherical figure of the ſpeculum, and to render its ſurface fit for the poliſher.

Fig. 3. An apparatus for examining the parabolic figure of the ſpeculum.

AA The mouth of the teleſcope, or edge of the great tube.

BB A thin piece of wood faſtened into, and fluſh with the end of the tube; to which is permanently fixed the annular piece of paſte-board cc, intended to cover, and to prevent the action of the correſponding part of the ſpeculum.

7   D. Ano-

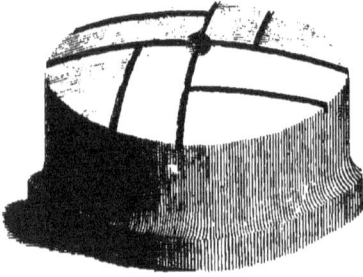

Fig. 1.ᵈ

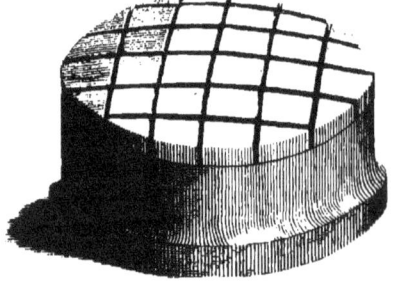

Fig. 2.ᵈ

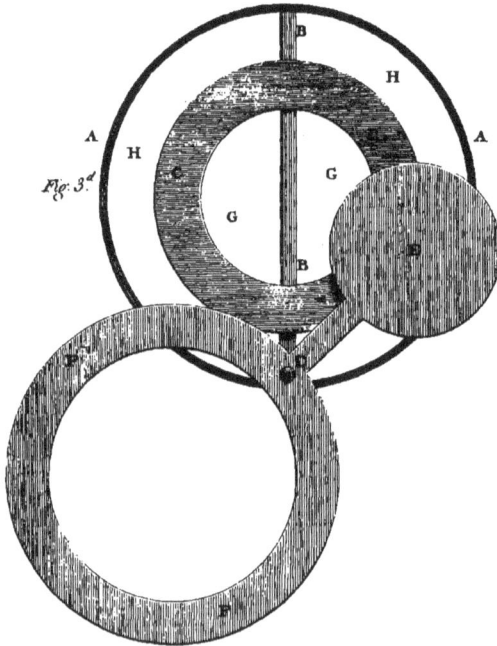

Fig. 3.ᵈ

D Another piece of paſte-board, fixed by a pin to the piece of wood BB, on which it turns as on a center; ſo that the great annular opening HH may be ſhut up by the riṅg FF, or the aperture GG by the imperforate piece E in ſuch manner that, in the firſt inſtance, the reflexion may be from the center, and in the latter from the circumference, of the great ſpeculum.